IMPACT CRATERS OF EARTH
WITH SELECTED CRATERS ELSEWHERE

Thomas Wm. Hamilton

Strategic Book Publishing and Rights Co.

Copyright © 2014 Thomas Wm. Hamilton. All rights reserved.

No part of this book may be reproduced or transmitted in any form or by any means, graphic, electronic, or mechanical, including photocopying, recording, taping, or by any information storage retrieval system, without the permission, in writing, of the publisher. For more information, send a letter to our Houston, TX address, Attention Subsidiary Rights Department, or email: support@sbpra.net.

Strategic Book Publishing and Rights Co.
12620 FM 1960, Suite A4-507
Houston, TX 77065
www.sbpra.com

For information about special discounts for bulk purchases, please contact Strategic Book Publishing and Rights Co. Special Sales, at bookorder@sbpra.net.

ISBN: 978-1-63135-353-6

Book Design: Suzanne Kelly

The term "crater" comes directly from Latin, where it was the word for cup. There is, in fact, an ancient constellation named Crater, depicting a fancy goblet sitting on the back of the constellation of Hydra, the sea serpent. It was Galileo Galilei (1564-1642) who adopted the word for our modern usage. His observations of the Moon with a telescope, starting November 30, 1609, revealed many cup shaped holes in the ground, so he gave them the obvious (to him) name. In recent times some people for whom this is not good enough have invented the term "astrobleme", coined from Latin words for star scar. I will stick with Galileo's usage. A few extremely large craters on other celestial bodies have been called "basins". I still prefer to stick with Galileo

The origin of craters was still being disputed as late as the 1960s, with some arguing that they were the open mouths of volcanoes (including an hypothesized but never seen category of "cryptovolcanoes"), while others insisted they were holes in the ground created when meteors, large rocks, fell from space. The Apollo landings on the Moon settled the issue. The Moon has a few small, dead volcanoes, but craters result from impacts by objects falling from space.

In this book we will look at all the known or suspected terrestrial craters, plus a sampling of interesting or typical craters from various planets and moons. Each crater is identified with a code letter and number, used in the indexes at the conclusion of the book. The code letters are A--Africa, B--Asia, C for Antarctica, D for Australia, E--Europe, N--North America, O--Oceana, S--South America. For other objects M--Mercury, L--Moon, R--Mars, T--Callisto, I--Mimas. In many cases I have added instructions on how to get to a particular crater and whether it has tourist facilities.

PARTS OF CRATERS

Many craters have a slightly raised *rim* around the edge. This is material tossed out when the crater was formed. There are also frequently *strew fields* (often called *rays* when seen on other worlds) of rocks, soil and rubble, which on low gravity/low atmospheric pressure places such as our Moon can extend for hundreds of miles. The *wall* of a crater is the part where it slopes down to the *floor*. The floor usually is rather smooth. In craters initially exceeding 16 miles diameter there can be a *central peak*. This is a hill, complex of hills, or even mountainous region in the center of a crater. It usually is the result of a splash effect when the crater is formed, as we shall see below, but we will also see a unique exception. The very largest craters seem to have few central peaks. Various forms of erosion can make some or all of these features either vanish or become buried.

In many craters rocks show *shock metamorphism,* a change in their physical structure from the effects of the impact. Coesite, or shocked quartz, is a good example, and often found with major craters. Some rocks are pulverized into a flour like appearance. Sometimes *shatter cones*, conical shaped rocks, are found, and are regarded as definite proof of a meteor impact.

SIZE OF CRATERS

Lunar samples retrieved on the Apollo missions showed tiny, almost microscopic craters, presumably caused by impacts from material no larger than a grain of sand. Nothing that small could survive falling through our atmosphere to make such a small crater on Earth. By contrast the largest crater known in the Solar System is on the Moon, and is 1300 miles across. It is named the Aitken South Polar Basin. Mercury's Caloris Basin is not much smaller. The largest on Mars is Hellas Basin, about 800 miles, while Callisto has Valhalla at about 1100 miles.

A WORD ON MINEROLOGY

While this is a work devoted to craters, a word or two regarding the objects creating them is appropriate. About 15% of meteorites are called siderolites, and are composed of up to 90% iron and 8% nickel. This is the type that responds to magnets.

Around 80% of meteorites are classified as chondrites, and are a mix of olivine $(MgFe)2SiO4$, pyroxene $(Si, Al)2O2$ with such other elements as Ca, Na, Fe, or Mg. Very small spherical inclusions called chondrules give this class their name.

Sulfides, silicates, and troilite (FeS) are also commonly found.

Other types include the carbonaceous and achondritic.

ORBITS

Spaceweather.com has recently enhanced its daily listings by showing the orbits of meteors observed entering our atmosphere. Most have perihelia close to Earth's orbit, and aphelia that may extend to the asteroid belt, or in some cases well beyond it. Craters are formed by objects for whom these orbits are fairly typical, even though most of the ones shown by Spaceweather never make it to the ground.

COMPLETENESS

New craters can be formed at any time. Chelyabinsk just missed getting a crater, as the destructive meteor landed in a pond. The only crater was a hole in the ice covering the pond. Old craters are discovered at the rate of one or two per year. This book covers all the terrestrial craters known at the time of writing.

EARTH'S CRATERS

AFRICA

A1   Amguid crater is located in Algeria's Tamanghasset district, over sixty miles from the nearest town, and extremely difficult to reach, as there are no roads in the region. It was discovered from the air by Roman Karpoff in 1948, although he did not get around to publicizing the discovery until 1953. The diameter is about 1250 feet, with a current depth close to 200 feet. However, wind blown sand and soil have filled it to an unknown depth. The impacting object seems to have come from the south southwest. It is estimated to be between ten thousand and one hundred thousand years old. No meteorite fragments have been recovered.
   Latitude 26^05' North; Longitude 04^24' East

Amguid Crater as seen from space.

A2   Aorounga Crater is located in the Ennedi district of Chad. Several different attempts at dating the crater have produced results ranging from 3500 to more than a million years old. Clearly the studies here are inadequate. The crater has a diameter of about 1.2 miles, with an outer ring diameter of 7 miles. Locals knew of this crater as a hole in the ground for centuries, but Europeans first learned of it in 1958.
   Latitude 19^06' North; Longitude 19^15' East

A3 Aouelloul crater is in Mauritania. It was first reported by Theodor Andre Monod in a 1951 publication. There is a small satellite crater to the north. It is fairly reliably dated to 3.1 million years ago, and has a diameter of about 1300 feet. The floor is covered to a depth of about 70 feet with wind blown sand and silt. The rim runs 45 to 60 feet high. Impact glass shards have been found in the area, and their analysis has aided the dating. This crater is privately owned.
   Latitude 20^15' North; Longitude 12^41' East

A4 Bosumtwi crater is located in Ghana, about 18 miles from the town of Kumasi. The crater is about 6.4 miles in diameter, and contains a lake about 1.25 miles in diameter. As Ghana's main lake, it is a popular resort location, although local beliefs regarding spiritual associations prohibit anything other than wooden boats on the lake. The lake is up to 1200 feet deep, but the floor of the crater lies 2400 feet below the surface. The crater is 1.07 million years old.
   This crater's lake is subject to extreme variations in size and depth depending on local rainfall, and is believed to have shrunk to a very small size during a prolonged drought in the Seventeenth Century. It is the only known crater in a vast area of west central Africa.
   Latitude 06^30' North; Longitude 01^25' West

A5 Gweni-Fada Crater is also sometimes called BP crater for the employer of the oil geologists who discovered it. It is located in Chad about 140 miles southeast of the Aorounga Crater (A2), and while they may have somewhat similar ages of under 345 million years, no direct relationship between them is known. The diameter is 9 miles.
   Latitude 17^25' North; longitude 21^45' East

A6 Kallkop Crater is in the Eastern Cape Province of South Africa, on a privately owned farm. The name comes from the Afrikaans language, "kalk" for limestone, and "kop" for head. The diameter is about 20,000 feet. It is about 250,000 years old.
   Latitude 32^43' South; Longitude 24^26' East.

A7 Kamil Crater is in the Uweinat desert of Egypt's New Valley Governate, just 0.4 mile north of Egypt's border with Sudan. It was discovered in 2008 from satellite photos by Vincenzo de Michele, and he led the first expedition to the site in 2010. It is 147 feet in diameter, 52 feet deep, and is believed to be less than 5000 years old. Estimates suggest an iron meteor 4.3 feet in diameter impacted. Initial weight would have been 20,000 pounds or more, and 1800 pounds of fragments ranging up to 183 pounds have been recovered.
   Latitude 22^01'06" North; Longitude 26^05'16" East

A8 Luizi Crater is in the Democratic Republic of Congo (that's the larger of the two countries using the Congo name), in the province of Katanga. While first reported by a German exploratory expedition in 1919, it remained unknown until visited and reported by Ludovic Ferriere in 2009. It is less than 575 million years old, since the rocks around it are that age, but a more precise age is not available. Satellite imagery suggests prolonged erosion. It is the first (and so far only) crater found in central Africa.
   Latitude 10^10' South; Longitude 28^00' East

A9   Morokweng Crater is in South Africa's North West Province, in the Kalahari Desert. It is 43 miles in diameter, but is completely buried, so it was detected by gravitational anomalies. Scientists drilling in the crater in 2006 retrieved a fragmentary piece of chondrite from the impacting object. The crater is believed to be about 145 million years old. The name comes from a neighboring village.
   Latitude 26^28' South; Longitude 23^32' East

A10   Oasis Crater is in Libya. It has a diameter of about eleven miles. It seems to be less than 120 million years old.
   Latitude 24^35' North; Longitude 24^24' East

A11   Ouarkziz Crater is in Algeria, near its border with Morocco, and was formerly called Tindouf Crater. The diameter is about two miles. It is less than 70 million years old.
   Latitude 29^00' North; Longitude 07^33' West

A12   Roter Kamm Crater is in Namibia, which is a former German colony, explaining why it has a German name meaning "red ridge". The crater is 50 miles north of the town of Oranjemund (another German based name). The diameter is 1.7 miles, with a depth of 430 feet. It is four to five million years old. It was first noticed as a possible crater by geologists in 1965, but not examined and confirmed as a crater until 1986. The impacter appears to have come from the southeast.
   This crater is in a restricted zone not open to outsiders, as it is in a diamond mining region.
   Latitude 27^46' South; Longitude 16^18' East

A13   Talemzane Crater is in Algeria, about 25 miles southeast of the small town of Hassi Delaa. It was first recognized as a crater in 1928, but serious studies had to wait until the 1950s. It is about 1.1 miles in diameter, and is believed to be less than three million years old.
   Latitude 33^19' North; Longitude 04^02' East

A14   Temimichat Crater is in Mauritania. It is about 0.45 mile in diameter, and is believed to be under 170 million years old. Few studies have been conducted.
   Latitude 24^15' North; Longitude 09^39' West

A15   Tenoumer Crater is in Mauritania, and is 1.1 mile in diameter. It is believed to be a relatively young 21,000 years old or less. It has 600 to 900 feet of mostly wind blown soil filling the floor, but is still 350 feet deep.
   Latitude 22^55' North; Longitude 10^24' West

A16   Tin Bider Crater is in a rugged, hilly region of Algeria. The diameter is 3.8 miles. It is under 70 million years old.
   Latitude 27^36' North; Longitude 05^07' East

A17   Vredevort Crater is in South Africa's Free State. Three towns, Vredevort, Parys and Koppies, are inside this 185 mile diameter crater. Parys has many tourist facilities with a Visitor Center, and the crater was declared a World Heritage Site in 2005. The crater was formed by the impact of an asteroid three to six miles across 2.023 billion years ago, so it is heavily eroded, with the Vaal River running along the outer rim. A central peak a couple thousand feet high still exists.
   Latitude 27^00' South; Longitude 27^30' East

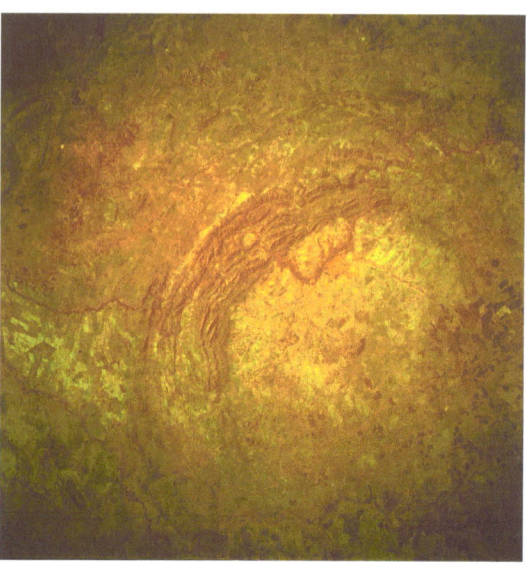

Vredevort Crater photographed on Mission 51 of the Space Shuttle. The Vaal River enters the crater on the upper left (northeast) and exits to the west (center right). The town of Vredevort is at the crater's center, with Parys at the loop in the river.

A18   Tswaing Crater is 25 miles north west of Pretoria, in South Africa. The diameter is 0.72 miles, and is about 310 feet deep. It is believed to be about 220,000 years old. The impacting object was a chondrite, based on studies of remnants. There is a museum at the site, which is easily reached from Pretoria via the M35 road north. The 190 foot crater floor has a small and very salty lake, which has long been mined for salt and other substances, and which animals have used as a salt lick.
   Latitude 25^40'10" South; Longitude 28^18'34.4" East

A19   Wembo-Nyama Crater is in the Eastern Kasai Province of Congo. It is about 2.5 miles in diameter, and about 60 million years old. The Unia River runs through the area. There is a central peak a few hundred feet high and 1900 feet across.
   Latitude 03^37'52" South; Longitude 24^31'07" East

A20   Arkenu Craters 1 &2 are in the midst of the Libyan desert. They are 4.2 and 6.4 miles in diameter respectively. They are estimated as under 140 million years old. Their status as impact craters is disputed, although shatter cones have been found.
   Latitude 22^04' North; Longitude 23^45' East

A21   Kgagodi Crater is in Botswana. The diameter is 2.2 miles, and it is estimated to be close to two billion years old.
   Latitude 22^28'3" South; Longitude 27^34'38" East

ASIA

B1 Beyenchime Crater is in Russia's Far Eastern province of Sakha, in the Yakutia region well north of the Arctic Circle. It has a diameter of 4.9 miles, and is believed to be about 40 million years old.
   Latitude 71^00' North; Longitude 121^40' East

B2 Bigach Crater is in Kazakhstan's Eastern province. It has a five mile diameter, and is about five million years old. The town of Novopavlovka is a few miles north.
   Latitude 48^34' North; Longitude 82^01' East

Bigach Crater and surroundings as photographed from the International Space Station in 2006.
The rectangular areas are farms.

B3 Chiyli Crater is also in Kazakhstan, 200 miles from the Aral Sea. The diameter is 3.8 miles, and it is 46 million years old.
   Latitude 49^10' North; Longitude 57^51' East

B4 Chukcha Crater is in Russia's Taimyr Peninsula, the frozen region near the Arctic Ocean. The diameter is 3.8 miles. It is less than 70 million years old.
   Latitude 75^42' North; Longitude 97^48' East.

B5 Dhala Crater is in India's Madhya Pradesh state. The inner crater is about 6.5 miles diameter, with an outer 15.5 mile ring. It is heavily eroded at an age of 1.8 billion years.
   Latitude 25^18' North; Longitude 78^08' East

B6 El'Gygytgyn Crater is in the Chuktoka region of Russia's Siberia. It is about 3.5 million years old. The 10.5 mile diameter crater has a 7.5 mile wide lake at its center, with an outlet via the Belaya River. The floor of the crater is buried under 4800 feet of sediment.
   Latitude 67^30' North; Longitude 172^05' East

B7 Jebel Waqf as Suwwan Crater is in Jordan's Ma'an region near the Saudi border. It is 3.8 miles in diameter and 50 million years old.
   Latitude 31^03' North; Longitude 36^48' East

B8 Kara Crater is in Russia's Yugoneki Peninsula, in the far north above the Arctic Circle. The crater is deeply buried, but drilling shows it may extend to a diameter as large as 75 miles. It is believed to be 70 million years old.
   Latitude 69^06'; Longitude 64^09' East

B9 Kara-Kul Crater is in the Pamir Mountains of Tajikistan, at one of the highest elevations of any crater, 12,800 feet above sea level. This remote crater was first identified from satellite photography. It has a diameter of 32.5 miles, and is partially filled with water. The eastern lake is 43 to 62 feet deep, the western part 725 to 755 feet deep. It is less than five million years old.
   Latitude 39^01' North; Longitude 73^27' East

Landsat captured this view of Kara-Kul in 2001, the first time anyone realized there was a crater here.

B10   Logancha Crater is in Russia's Siberia. It is buried, with a village of the same name atop it. The diameter is 12.5 miles. It is about 40 million years old.
   Latitude 65^31' North; Longitude 95^56' East

B11   Lonar Crater is in the Maharashtra state of India. It contains a saline lake 3900 feet in diameter and up to 450 feet deep within the 1.3 mile diameter crater. The impacting object struck about 570,000 years ago, coming in from the east at an angle of 35 to 40 degrees. The surroundings include a town of the same name with various historically important Hindu temples and resort hotels. At one time the area was commercially important for production of salt and carbonates of soda.
   Latitude 19^58'36" North; Longitude 76^30'30" East

Lonar Crater photographed by NASA's Terra satellite in 2004.

B12   Macha Crater in Russia's far eastern Sakha province is actually a complex of five craters ranging from 190 to 1600 feet in diameter. They are believed to have been formed 7300 years ago when an iron meteorite exploded just before hitting the ground. The two largest craters combine to form Abram Lake. To their north is a 65 foot deep crater with no water in it. Carbon dating on burned logs performed in 1984 provides a reasonably reliable age.
   Latitude 60^06'; Longitude 117^35' East

B13   Popigai Crater is located in Siberia, 180 miles from the town of Khatanga. The impact was 35.7 million years ago. A chondrite created a crater 55 miles in diameter. The crater tends to be snow covered 250 to 270 days per year with the lowest recorded temperature -63F, but for a long time visitors were prohibited anyway because the impact's pressure turned graphite in the ground into small, industrial quality diamonds. The Russian government estimates there are enough of these present to supply all of Earth's needs for industrial diamonds for 3500 years.
  Latitude 71^39' North; Longitude 111^11' East

B14   Ragozinka Crater is in Russia. It is completely buried, but has been verified through drilling rock samples. It is 5.5 miles in diameter, and 46 million years old.
  Latitude 58^44' North; Longitude 61^48' East

B15   Shiva Crater is located in the Indian Ocean, off the western coast of India. Its status as an impact crater is strongly disputed. It is 350 by 270 miles, and supposedly formed 65 million years ago by a 25 mile diameter object coming in at a low angle. Dredging has turned up melted rock, shocked quartz, and Fe_2O_3 with infusions of iridium, all of which are cited as evidence of an actual impact.
  Latitude 18^40' North; Longitude 70^14' East

B16   Shunak Crater is in the southeast of Kazakhstan's Qaraghandy Province. It is 1.8 miles in diameter, and has a low rim with a flat floor. It is about 45 million years old.
  Latitude 47^12' North; Longitude 72^42' East

B17   Sobolev Crater is in Russia, near its Pacific coast. It is 160 feet across, and is believed to be less than a thousand years old.
  Latitude 46^18' North; Longitude 137^52' East

B18   Tabun-Khara Crater is in southeastern Mongolia, Dornogovi Province. It is about one mile in diameter, and believed to be around 150 million years old. The rim is 70 to 100 feet above the floor, which is filled to a depth of 600 feet with lake deposits, although the crater is dry today. It was first identified as a crater in the 1960s. Possible smaller associated craters may exist 25 and 45 miles to the northeast.
  Latitude 44^07'52" North; Longitude 109^36'14" East

B19   Umm al Bini Crater is in southern Iraq, in the Maysan District's Central Marshes. At one time it was a lake, but this has pretty much dried up. The diameter is 2 miles, and it is believed to have formed not more than 5000 years ago, occasioning legends of terror and fire falling from the sky. The crater is about 10 feet deep today. Hazardous conditions due to warfare have limited the studies of this crater to examination of orbital photography.
  Latitude 31^14'29" North; Longitude 47^06'21" East

B20 Wabar Crater is in Saudi Arabia's Empty Quarter, which is even less hospitable than those far northern Russian locations. It seems to have resulted from a fire ball seen over Yemen on September 1, 1704. On the ground, it was discovered in 1932 by a team exploring for oil. There are several craters, possibly including two buried in drifting sand. The exposed craters are 360, 150, and 40 feet in diameter. The area has FeNi fragments from what seems to have been an exploded siderolite, and coesite has been found.
    Latitude 21^30' North; Longitude 50^28' East

B21 Xiuyuan Crater is in China's Sichuan Province. The diameter is 1.1 miles. It is more than 50,000 years old. There are tourist accomodations nearby, mainly geared to visitors at several historical sites unrelated to the crater.
    Latitude 40^21' North; Longitude 123^27' East

B22 Zhamanshin Crater is in Kazakhstan. It is 8.8 miles in diameter. A Russian language website says that there are tourist facilities available there.
    Latitude 48^24' North; Longitude 60^58' East

B23 Sikhote-Alin Craters are in eastern Siberia. The 158 mostly small craters result from a meteor that exploded shortly before impact on February 12, 1947. Many museums around the world have siderolite fragments from this event.
    Latitude 46^09'05"N; Longitude 134^39'28' East

ANTARCTICA

Antarctic craters are unconfirmed, and little has been learned of the few suspected, since they are buried beneath a couple thousand feet of snow and ice in a region where it is not easy to bring in specialized equipment.

C1   Bowers Crater is estimated to be 62 miles in diameter. The age is undetermined.
   Latitude 71°12' South; Longitude 178° East

C2   Ross Crater is believed to be one of Earth's largest, with a diameter of 350 miles. It is more than 24 million years old.
   Latitude 77°30' South; Longitude 179°30' East

C3   Wilkes Crater is in Wilkes Land, with a 310 mile diameter, and less than 500 million years old.
   Latitude 70° South; Longitude 120° East

AUSTRALIA

D1   Acraman Crater is in South Australia. It has a diameter of 12.5 miles, and is 580 million years old. It contains a seasonal lake. The crater was discovered in 1986. Ejecta from the impacting chondrite has been found 180 miles away in the Flinders Mountain Range. Shatter cones, coesite and iridium deposits suggest that this crater could have been fifty miles in diameter before erosion.
   Latitude 32^01' South; Longitude 135^27' East

D2   Amelia Creek Crater is in Australia's Northern Territory, and is not to be confused with Amelia "Amy" Pond. The diameter is 12 miles. It is 1.6 billion years old. Shatter cones have been found.
   Latitude 20^55' South; Longitude 134^50' East

D3   Boxhole Crater is in the Northern Territory. The diameter is about 520 feet, and it is 54,000 years old. It is 110 miles northeast of Alice Springs, 160 miles driving. It was discovered in 1937. The impacting object was a siderolite. The name derives from the sheep ranch ("station" in Australian parlance) where it is located.
   Latitude 22^35'45" South; Longitude 135^11'43" East

D4   Connolly Basin Crater is in West Australia's Gibson Desert. It is 28 miles west of Windy Corner, and adjacent to Talawana Trail. The diameter is 5.7 miles, and it is less than 60 million years old. This crater was discovered in 1985.
   Latitude 23^32'18" South; Longitude 124^45'25" East

D5   Crawford Crater is in South Australia, near the city of Adelaide. It is 5.4 miles in diameter, and more than 35 million years old. Coesite has been found around the crater.
   Latitude 34^43' South; Longitude 139^02' East

D6   Dalgaranga Crater is in West Australia, 45 miles west of Mount Magnet, and 62 miles north of Yalgoo. The crater has a diameter of 80 feet, and is ten feet deep, with evidence it came in from the south southeast. It is 270,000 years old, and was discovered in 1921. Fragments of the stony iron impactor are in a local museum.
   Latitude 27^38'06" South; Longitude 117^17'20" East

D7   Darwin Crater is in the western part of Tasmania, 16 miles south of the town of Queenstown, and inside the Franklin-Gordon Wild Rivers National Park. it is 0.75 mile in diameter, and 816,000 years old. Coesite has been found.
   Latitude 42^18'15" South; Longitude 145^39'27" East

D8   Flaxman Crater is in South Australia, northeast of the state capitol of Adelaide. The diameter is 6.2 miles, and the age is more than 35 million years. Shatter cones have been found.
   Latitude 34^37' South; Longitude 139^04' East

D9   Foelsche Crater is in Australia's Northern Territory. The diameter is 3.7 miles, and the age at least 545 million years. The crater is 5.3 miles southeast of the town of Borroloola, and shares its name with a river. Coesite has been found.
   Latitude 16^40' South; Longitude 136^47' East

D10   Glikson Crater is in West Australia, and is named for one of Australia's leading scientists studying craters. It is in the Little Sandy Desert. This crater is 12 miles in diameter, and less than 508 million years old. It was identified in 1997 upon the discovery of shatter cones.
   Latitude 23^59' South; Longitude 121^34' East

D11   Goat Paddock Crater really exists despite the name, and is in West Australia 68 miles west southwest of the town of Halls Creek. It is 3.1 miles in diameter, and less than 50 million years old. Shatter cones and coesite have been found.
   Latitude 18^20' South; Longitude 126^40' East

D12   Gosses Bluff Crater is in the Northern Territory 109 miles west of Alice Springs, from which day trips are easily arranged. Originally 14 miles in diameter, it has been eroded down to little more than three miles. It is 142.5 million years old. The crater was first detected while drilling for oil, when shatter cones were found.
   Latitude 23^49'15" South; Longitude 132^18'28" East

D13   Goyder Crater is in the Northern Territory's Arnhem Land, next to the Goyder River. The crater is 1.8 miles in diameter and under 1.4 billion years old. Shatter cones and coesite have been found.
   Latitude 13^28' South; Longitude 135^02' East

D14   Henbury Crater is actually a complex of 12 craters in the Northern Territory 90 miles southwest of Alice Springs, reached via the Stuart Highway. The largest crater has a diameter of 600 feet, the smallest is 20 feet across. They were formed 4200 years ago by a siderolite which apparently shattered in an explosion before reaching the ground. Troilite has been found. A public camp ground adjoins the craters, but those using it are warned to bring their own water.
   Latitude 24^34'19" South; Longitude 133^08'54" East

D15   Hickman Crater is in West Australia, 21 miles north of the town of Newsom. The diameter is 780 feet, it is 100 feet deep, and less than 100,000 years old. The meteor is believed to have hit the Earth coming in at a high angle from the south. It was identified in 2008.
   Latitude 23^02'14" South; Longitude 119^40'59" East

D16   Kelly West Crater is in the Northern Territory near Tennant Creek. The diameter is 4.2 miles. It is more than 550 million years old. Shatter cones identifying it as a crater were found in 1973.
   Latitude 19^56' South; Longitude 133^57' East

D17   Lawn Hill Crater is in Queensland, 130 miles north northwest of Mount Isa. This siderolite impact left an 11 mile crater now 509 million years old. It was identified in 1987 when shatter cones and coesite were found. Evidence suggests it was water filled shortly after impact. It is dry today.
    Latitude 18^40' South; Longitude 138^39' East

D18   Liverpool Crater is in the Northern Territory's Arnhem Land, adjacent to the Liverpool River. It is a mile in diameter, and 150 million years old. It was identified in 1970 following the discovery of coesite. The impacting object seems to have come in from the southwest.
    Latitude 12^24' South; Longitude 134^03' East

D19   Matt Wilson Crater is in the Northern Territory 10 miles north of the Victoria River Roadhouse, seventy miles from the town of Katherine, and in the Judbarra/Gregory National Forest. It is 4.5 by 4 miles in diameter, and 1.4 billion years old. The impacting object may have come in from the northeast.
    Latitude 15^30'04" South; Longitude 131^10'43" East

D20   Mount Timondina Crater is in South Australia, 28 miles south of the town of Oodnadatta. The diameter is 2.5 miles, and it is less than 110 million years old. It was identified in 1976.
    Latitude 227^56'40" South; Longitude 135^21'30" East

D21   Piccaninny Crater is in West Australia near a creek of the same name in Piccnululu National Park. It has a 4 mile diameter, and is less than 360 million years old. It was identified in 1983.
    Latitude 17^25'30" South; Longitude 128^26'10" East

D22   Shoemaker Crater is in West Australia, 62 miles northwest of the town of Wiluna. It is named for the prominent American astronomer Eugene Shoemaker, who was killed in an auto accident in Australia. (He also has a crater on the asteroid Eros named for him.) Previously the crater shared a name with nearby Lake Teague. It is 19 miles in diameter and 1.63 billion years old. Shatter cones and coesite led to its being recognized in 1974.
    Latitude 29 ^52' South; Longitude 120^53' East

D23   Spider Crater is in West Australia twelve miles east of Mount Burnett Roadhouse on Gibb River Road. The diameter is 8 miles, and the age more than 570 million years. Shatter cones have been found.
    Latitude 16^44'22" South; Longitude 126^05'25" East

D24   Strangways Crater is in the Northern Territory near a river of the same name. It is just under ten miles in diameter, but prior to erosion may have been as much as 20 miles across. It is 646 million years old. The impacting object was achondritic. Shatter cones and coesite led to the crater's discovery in 1971.
    Latitude 15^12' South; Longitude 133^35' East

D25   Tookoonooka Crater is in Queensland near the town of Durham Downs. It is 34 miles in diameter, and 128 million years old. It was discovered in 1989 when oil drilling turned up coesite. Seismic readings suggest a second, smaller crater called Talundilly lies buried nearby, but absent any drilling this has not been verified.
    Latitude 27^07' South; Longitude 142^50' East

D26   Veevers Crater is in West Australia's Gibson Desert, pretty much in the middle of nowhere. It is 250 feet across, with a five foot rim, and 25 feet deep. It is less than 20,000 years old. Fragments found since 1975 show the impact came from an unusual type of siderolite, with rare minerals such as octahedrite and schreibersite.
   Latitude 22^58' South; Longitude 125^22' East

D27   Wolfe Creek Crater is in West Australia 93 miles south of the town of Halls Creek via the Tanami Road. Today it is protected in the Wolfe Creek Meteorite Crater National Park. The park is home to such Australian wildlife as red kangaroos, macaws, and brown ringtail dragons. That last may have been the reason the crater was featured in a 2005 horror film. The crater was identified in 1947 as caused by a siderolite. The crater is 2500 feet across, 190 feet deep, and 5300 years old.
   Latitude 19^10'18" South; Longitude 127^47'44" East

D28   Woodleigh Crater is in West Australia northwest of the town of Ajana. It has a 25 mile diameter, and is 364 million years old. This crater is buried beneath sedimentary rocks.
   Latitude 26^03' South; Longitude 114^39' East

D29   Yarrabubbo Crater is in West Australia between the towns of Sandstone and Meekatharra. The diameter is 19 miles, and it is 1.13 billion years old. It was identified in 2003 from the discovery of shatter cones and coesite.
   Latitude 27^10' South; Longitude 118^50' East

EUROPE

E1   Azuara Crater is in Spain, 30 miles south of Zaragoza. It has a 24 mile diameter, and is 40 million years old. It is named for a village within the crater. Most Spanish geologists deny this is a meteor crater, but coesite and other indicators have been found. It was identified in 1980.
   Latitude 41^10' North; Longitude 0^55' West

E2   Boltysh Crater is in Kirovohrad, Ukraine. The crater is buried, with the town of similar name sitting atop it, near the Tiasmyn River. The crater's diameter is 15 miles, with breccia found covering 2500 square miles. There is a central peak buried beneath 1600 feet of sediment. This crater is 65.1 million years old.
   Latitude 48^45' North; Longitude 32^10' East

E3   Dellen Crater is a popular Swedish resort lake, with trout fishing prominent. The lake is about 12 miles across, with a peninsula nearly dividing it into two separate lakes. The nearest town is Nasviken. The crater is about 89 million years old, and its lake is so popular that Uppsala University named asteroid 7704 Dellen.
   Latitude 61^51' North; Longitude 16^42' East

E4   Dobele Crater is in Latvia, buried beneath a town of the same name. The crater diameter is 2.7 miles, and it is believed to be 290 million years old.
   Latitude 56^35' North; Longitude 23^15 East

E5   Gardnos Crater is near the Norwegian town of Nes. There is a paved highway through the crater, as well as two hiking trails through it, and Norway advertises it as one of the easiest craters to visit. The diameter is three miles, and it is about 500 million years old.
   Latitude 60^39' North; Longitude 09^00' East

Entrance to Gardnos Crater.

E6   Granby Crater is in Sweden. It is buried, but drilling has turned up the usual geological indicators of a crater. The diameter is 1.8 miles, and it is about 470 million years old.
   Latitude 58^25' North; Longitude 14^56' East

E7   Gusev Crater exists in Russia as well as on Mars. The Russian one is near Rostov. It is about 1.8 miles in diameter, but buried. It is around 49 million years old. This may have been formed as part of a pair of impacting objects with Kamensk Crater (E9).
   Latitude 48^26' North; Longitude 40^32' East

E8   Janisjarvi Crater is in Russia. It is 8.6 miles in diameter and 700 million years old.
   Latitude 61^58' North; Longitude 30^55' East

Water-filled Jänisjärvi Crater photographed by NASA's Landsat in 1999.

E9   Kamensk Crater is in Russia. It has a sixteen mile diameter, but is buried. It is around 49 million years old. This may have been the primary in a pair of objects that formed this and Gusov Crater (E7).
   Latitude 48^21 North; Longitude 40^30' East

E10   Kaluga Crater is in Russia. It has a 9 mile diameter, and is 380 million years old. This crater is buried.
   Latitude 54^30' North; Longitude 36^12' East

E11   Iceland is not a crater, but the amount of iridium found in its rocks has been one of the factors leading to suggestions that it may have been formed at the same time as Chicxulub Crater (N11) in Mexico, presumably by a satellite of that impact object. If correct, this impact would have hit right on the boundary between the North American and European tectonic plates, where Earth's crust was thin, breaking through to the mantle.

E12   Ilumetsa Crater is in Estonia's Polava County. The diameter is 250 feet. It is located in a pine and birch forest just over one thousand feet from a major highway which is marked with a sign that commemorates a local legend connecting the crater to the devil. It is about 6600 years old.
   Latitude 57^57'36" North; Longitude 27^24'10" East

E13   Ilyinets Crater is in Ukraine, buried beneath the village of Lugove, between the Sob and Sobyk Rivers. The diameter is five miles, and it is 378 million years old. It was first studied in 1851, and reported as a crater as early as 1898.
   Latitude 49^07' North; Longitude 29^06' East

E14   Iso-Naakkima Crater is in Finland six miles from the town of Pieksamaki. It was recognized as a buried crater in 1989. The diameter is 1.8 miles, and the age is believed to be about 1.2 billion years. A small lake sits above part of the crater, but is unrelated.
   Latitude 62^11' North; Latitude 27^09' East

E15   Kaalijarv Crater is in Estonia on Saaremaa Island. The diameter is 420 feet, with a lake ("jarv" is Estonian for lake) 75 feet deep, and 20 feet of sediment. This is a complex of nine craters, the smaller ones with diameters ranging from 40 to 140 feet, and with depths of 3 to 14 feet. The impact is believed to have been something over 4000 years ago, and with vegetation scorched out to four miles, it is virtually certain human fatalities accompanied this impact. It subsequently became a sacred lake, surrounded by a wall, and with animal sacrifices performed. It is also suspected of being the historical event at the root of several myths. There is a visitor center, and asteroid 4227 is named Kaali for the crater.
   Latitude 58^24' North; Longitude 22^40' East

E16   Kardla Crater is in Estonia at the town of Tonui on the island of Hiiumaa. It is 4.5 miles in diameter, and 455 million years old. It is buried.
   Latitude 59^01' North; Longitude 22^46' East

E17   Karikkoselka Crater is in Finland, slightly southwest of the town of Petajavesi. It is just under a mile in diameter, and 230 million years old. It is buried about 480 feet deep, but drilling has turned up shatter cones and other indicators.
   Latitude 62^13' North; Longitude 25^15' East

E18   Keurusselka Crater is in Finland between the towns of Keuruu and Mantta. It has a diameter of 19 miles, and contains a lake up to 130 feet deep. It is believed to be 1.8 billion years old. It was identified as a crater in 2003.
   Latitude 62^08' North; Longitude 26^34' East

E19 Lappajarvi Crater is in Sweden between the towns of Lappajarvi and Vimpeli. It has a fourteen mile diameter, and is about 76.2 million years old. It holds a lake, with the crater's central peak the island of Karnansaari.
    Latitude 63^12' North; Longitude 23^49' East

E20 Lockne Crater is in Sweden thirteen miles south of the city of Ostersund. The crater's diameter is 4.6 miles, and it has been fairly accurately dated as 458 million years old thanks to extensive findings of recognizable fossils. These fossils all indicate the impact came in a shallow sea bed. This may be part of a pair of impacts, the mate being Malingen Crater (E23) ten miles away.
    Latitude 63^00' North; Longitude 14^49' East

E21 Logoisk Crater is in Belarus, buried beneath the city of Lahojsk, which offers tourist facilities as a popular winter ski resort. The crater is 9.3 miles in diameter, and about one billion years old.
    Latitude 54^12' North; Longitude 21^48' East

E22 Lumparn Crater is in Finland, forming a bay in the island of Aland. It is six miles in diameter and estimated to be one billion years old. It was first studied as a crater in 1979, and confirmed in 1993.
    Latitude 60^09' North; Longitude 20^06' East

E23 Malingen Crater is in Sweden, ten miles from Lockne Crater (E20). It is believed to have been a formed by a satellite of the object that formed Lockne. Malingen is just under half a mile in diameter, and 458 million years old. It probably impacted into a shallow sea. Coesite has been found.
    Latitude 62^55' North; Longitude 14^45' East

E24 Mien Crater is in Sweden, 7.5 miles southwest of the town of Tingsryd. The crater has a diameter of six miles, but contains a lake 3.4 miles across. The crater is 121 million years old. The Drevan River flows into it, with outflow from the Miean River.
    Latitude 56^25' North; 14^52' East

E25 Mizarai Crater is in Lithuania, and is named for a village located inside the crater, not for the star Mizar. The diameter is 3.1 miles, and the age is 520 million years. It was discovered during a seismic survey, being totally buried.
    Latitude 54^01'; Longitude 23^54' East

E26 Mjolnir Crater is in the Barents Sea between the north coast of Norway and the Svalbard Islands. It is just east of Bjornoya Island. It is about 25 miles in diameter, and 142 million years old. There are two rings and a central peak whose top seems to have been eroded by sea action. In any case the whole thing is under water. Shocked quartz and iridium enrichment confirm the identification. The impact seems to have created a tsunami that left traces on Greenland, Sweden, Novaya Zemlya, and possibly France.
    Latitude 73^48' North; Longitude 29^40' East

E27   Morasko Crater is in Poland, near the northern edge of the city of Poznan. The diameter is 450 feet, and it is about 5000 years old. It is on the slope of a glacial moraine, providing additional grounds for geological study. There is actually a complex of seven craters, the first of which was discovered by occupying German troops in 1914. Several large meteoritic fragments have been found, the largest was found in 2006 and weighs over a metric tonne. They are FeNi with pyroxenes. The site is served by two bus lines from downtown Poznan.
    Latitude 52^29' North; Longitude 16^54' East

E28   Neugrund Crater is in the sea near Estonia's island of Osmussaar, which can be reached by boat from Dirham. It has a five mile diameter, and is 470 million years old.
    Latitude 59^20' North; Longitude 23^40' East

E29   Obolon Crater is in Ukraine, 120 miles from the capitol of Kiev, near Poltova. It is 12.5 miles in diameter, and 169 million years old. Chlorine in the material comprising the crater has been interpreted as suggesting it landed in a shallow sea. Today it is buried.
    Latitude 49^35' North; Longitude 32^55' East

E30   Paasselka Crater is in Finland's Southern Savouin Province. The name means "stone lake", and this 6.2 mile wide crater is water filled. It is 250 feet deep. The age is about 231 million years. Drilling has turned up the usual indicators of a crater.
    Latitude 62^02' North; Longitude 29^05' East

E31   Ries Crater is 3.7 miles north east of Nordlingen, Bavaria. The diameter is 14 miles. The impacting object appears to have come in at a 30 to 50 degree angle from the west southwest 15.1 million years ago. Rubble was strewn as far as 280 miles away in Bohemia and Moravia. A probable satellite impact occurred 26 miles to the west southwest at Steinheim (E35).
    Latitude 48^53 North; Longitude 10^37 East

E32   Ritland Crater is in Norway, on a farm of the same name in Rogaland County. The diameter is 1.3 miles, and the age 520 million years. It was first detected in 2000, and is 1150 feet deep.
    Latitude 58^14'48"; Longitude 06^25'18"

E33   Rochechouart Crater is in France's department of Haut Vienne, with the village of La Judie sitting inside the crater. It is 13.9 miles in diameter, and 201 million years old.
    Latitude 45^50'; Longitude 00^56' East

E34   Rubielos de la Cerida Crater is in Spain, named for a village in Aragon that lies within the northern part of the crater. It is 27 miles in diameter, and 32 million years old. Coesite and shatter cones have been found.
    Latitude 40^49' North; Longitude 01^27 East

E35   Steinheim Crater is in the Swabian Alps of Baden-Wurttemberg. The name is taken from the town inside the crater, Steinheim am Albuch. It is 2.35 miles in diameter, and 15 million years old. Shatter cones and sediments from an ancient lake that once existed inside the crater have been found. This may have been formed by a satellite of the object that created the Ries Crater, E31.
    Latitude 48^41' North; Longitude 10^04' East

E36   Suvasvesi Crater North and South are a pair in Finland, separated only by some small rocky islands. The north crater sits in Kuukkarinselka lake, 2.2 miles in diameter, and up to 292 feet deep. The south crater has a similar size, and is in Haapaselka lake. Both are about 710 million years old. Shatter cones were found in 2001.
    Latitude 62^42' North; Longitude 28^10' East

E37   Ternovka Crater is in Ukraine. It is buried, but drilling suggests a diameter of 6.8 miles, and an age of 280 million years. It is near the city of Globino.
    Latitude 48^08' North; Longitude 33^31' East

E38   Tvaren Crater is in Sweden. It is buried, with a diameter around 1.2 miles, and an age of 455 million years. It is near the town of Nykoping.
    Latitude 58^46' North; Longitude 17^25' East

E39   Ullapool Crater is in Scotland, named for the village on its western edge, as the crater extends beneath a body of water known as The Minch. It is 8 miles in diameter, and 1.18 billion years old. Ejecta has been found more than 35 miles away. It was recognized as a crater in 2008.
    Latitude 58^06' North; Longitude 05^55' West

E40   Vepriai Crater is in Lithuania, with a diameter of 5 miles. It is believed to be over 160 million years old, and is buried. It is between the capital of Vilnius and the town of Ukmerge.
    Latitude 55^05' North; Longitude 24^35'

E41   Zapadnaya Crater is in Ukraine. It is 2 miles in diameter, 165 million years old, and buried. It is near the town of Skvira.
    Latitude 49^44' North; Longitude 29^00' East

E42   Zeleny Gai Crater is in Ukraine, another buried crater. The diameter is 2.3 miles, and it may be 80 million years old. It is between the town of Bobrinets and the city of Krivoy.
    Latitude 48^04' North; Longitude 32^45' East

NORTH AMERICA

N1   Ames Crater is in Oklahoma, buried two miles north of the town of Ames. The diameter is 10 miles, and it is about 470 million years old. The impacting object was a chondrite. This may have been part of a group that includes the Decorah (N17), Slate Islands (N56), and Rock Elm (N50) craters formed by related impacts.
   Latitude 36^15' North; Longitude 98^12 West

N2   Avak Crater is in Alaska, 7.5 miles south of the city of Barrow. The diameter is 5 miles, and may be 50 million years old. This buried crater was first detected in 1951 during a drilling for oil.
   Latitude 71^15' North; Longitude 156^30' West

N3   Barringer Crater is in Arizona, 37 miles east of Flagstaff, and 18 miles west of Winslow. It can be reached from exit 233 off Interstate 40. This crater is 3900 feet wide, with an age that is disputed, a maximum of fifty thousand years, possibly half that. It is on the Coconino Plateau (a reddish sandstone) 5710 feet above sea level. It is 570 feet deep, with 700 feet of rubble beneath. The impacting object was a siderolite, and to this day fragments of nickel-iron are found in the area. This crater has parking, view stands, and other tourist facilities. It was known to local Native Americans, but first reported by an exploring Army patrol following the Civil War as a volcanic structure. Daniel Barringer, a mining engineer, correctly identified it, and claimed it for his family, which still owns the crater.
   Latitude 35^01'33" North; Longitude 111^01'21" West

N4   Beaverhead Crater is in Montana. The diameter of the buried crater has been measured as 37 miles. It is 600 million years old. It was discovered in 1990 through finding shatter cones in the area. It is in the Beaverhead Mountains, near the border with Idaho. The nearest towns are Grant to the north and Dell to the east.
   Latitude 44^36' North; Longitude 113^12' West

N5   Brent Crater is in Ontario, buried beneath the township of Deacon. The diameter is 2.4 miles, It is about 453 million years old. The impacting object was a chondrite. The crater was discovered in 1951 by drilling. It can be reached by taking Brent Road south from Ontario Highway 17. The crater contains two small lakes, Gilmour on the east and Tecumseh on the west. There are also two hiking trails through the crater.
   Latitude 41^04'22" North; Longitude 78^28'38" West

N6   Brushy Creek Crater is in Louisiana, 5.8 miles southwest of the county seat of Greensburg. It is 1.2 miles in diameter, and less than 1.9 million years old. Highway 37 cuts through the crater. It was discovered in 1997 as a by product of a search for oil. Coesite has been found.
   Latitude 30^46' North; Longitude 90^45' West

N7   Calvin Crater is in Michigan between the towns of Marcellus and Glenwood. Buried, it was discovered by crews drilling for oil in 1987. It is 7.6 miles in diameter, and about 450 million years old.
   Latitude 41^50' North; Longitude 85^57' West

N8   Carswell Crater is near Saskatchewan's northwestern corner. There are no nearby towns. La Loche is about a hundred miles south, but Saskatchewan Highway 55 does provide access, as there was at one time a uranium mine in the area. The crater is 25 miles in diameter, with traces as much as 77 miles out from the center. The Douglas River runs along the outer edge of the crater rim. Shatter cones led to the discovery of the crater in 1961. It is about 115 million years old.
   Latitude 58^26'26" North; Longitude 109^39'24" West

N9   Charlevoix Crater is on the north shore of the St. Lawrence River in Quebec, with the town of Eboulements inside it. The diameter is 34 miles. It is 342 million years old. The impacting object was a chondrite. The crater was recognized in 1965 through the discovery of shatter cones. Route 362 provides easy access.
   Latitude 47^32' North; Longitude 70^18' West

N10   Chesapeake Bay Crater is located inside Virginia's part of Chesapeake Bay. The diameter is 55 miles, but ejecta from the impact was found as far away as Atlantic City, NJ, a couple hundred miles north in 1983. It is 35.5 million years old, and fracturing of the underlying rocks goes as deep as seven miles. The depression is still being filled as coastal land gradually slips into it.
   Latitude 37^17' North; Longitude 76^01' West

N11   Chicxulub Crater is on the northwestern edge of the Yucatan Peninsula, extending into the Gulf of Mexico. The name is Mayan, and is taken from a village inside the crater. It is pronounced chick shuh luhb. The crater's diameter is 190 miles, and although buried, extends 12 miles deep. It was formed 64.98 million years ago, and as with so many other craters was first detected during a search for oil. The outer rim in the Yucatan is defined by a series of cenotes, water filled pits in limestone which in pre-conquest times were used for sacrifices of humans, animals, and precious objects. Coesite and fragments of a carbonaceous chondrite make this an identified crater, although buried. It is, of course, famous as the impact that wiped out the dinosaurs, pterosaurs, and many other species. The impacter is believed to have been something over six miles in diameter, and may have had a satellite that struck where Iceland is today.
   Latitude 21^24' North; Longitude 89^31' West

N12   Clearwater Lakes East & West are in Quebec near Fort McKenzie. The eastern lake is 19 miles in diameter, the western is 22 miles. They were formed simultaneously 290 million years ago. The western lake has an island central peak, while the eastern's peak is submerged. My letter to *Sky & Telescope* magazine in 1978 was the first to cite this pair as evidence asteroids could have moons.
   Latitude 56^05' and 56^13' North; Longitude 74^07' and 74^30' West

N13   Cloud Creek Crater is in Wyoming, 48 miles northwest of Casper. It is 4.3 miles in diameter and 190 million years old. This buried crater was another one discovered through drilling for oil.
   Latitude 43^07' North; Longitude 106^45' West

N14   Couture Crater is in northern Quebec, east of the town of Povungnituk. It has a diameter of 5 miles, and is 430 million years old. Lac du Couture fills the crater.
   Latitude 60^05'36" North; Longitude 75^19'43" West

N15   Crooked Creek Crater is in Missouri. It is 4.3 miles in diameter, and 320 million years old. It is easily accessed as local road VV off Highway 19 goes right through the crater.
   Latitude 37^50'08" North; Longitude 91^23'42" West

N16   Decaturville Crater is in Missouri, sixteen miles north of the town of Lebanon. Highway 5 runs along the side of the crater, which is 3.8 miles in diameter and less than 300 million years old.
   Latitude 37^54' North; Longitude 92^43' West

N17   Decorah Crater is in Iowa, on the east side of the town it is named for. The diameter is 3.5 miles, and the age 470 million years. State highway 9 and U.S. Route 52 provide easy access, although this crater is well buried. There are suggestions that Ames (N1), Rock Elm (N50), Slate Islands (N56) and this crater were formed by related impacts.
   Latitude 43^18'50" North; Longitude 91^46'20" West

N18   Deep Bay Crater is in Saskatchewan at the south end of Reindeer Lake. It is 700 feet deep, although the rest of the lake averages only 65 to 70 feet deep. Coesite has been found around the bay. It is eight miles in diameter, and 99 million years old.
   Latitude 56^24' North; Longitude 102^59' West

N19   Des Plaines Crater is in Illinois, buried by glacial till beneath the town sharing its name. Shatter cones have been found. It is 5.5 miles in diameter, and less than 280 million years old.
   Latitude 42^03' North; Longitude 87^52' West

N20   Eagle Butte Crater is in Alberta northwest of the town of Orion. It is 6.2 miles in diameter, and less than 65 million years old. Shatter cones and a central peak prove its formation and age. Highway 41 runs through the crater 22 miles south of Medicine Hat, and it is just west of Cypress Hills National Park.
   Latitude 49^42' North; Longitude 110^30' West

N21   Elbow Crater is in Saskatchewan just north of the village of Riverhurst, and near Lake Diefenbaker (named for a former Canadian Prime Minister). The crater is five miles in diameter and 395 million years old. The odd name refers to a kink in a local river caused by the crater.
   Latitude 50^59' North; Longitude 106^43' West

N22   Flynn Creek Crater is in Tennessee five miles south of Gainesville. It is 2.2 miles in diameter and 360 million years old. The area was a sea when the impact happened, as fossil conodonts, an extinct shelled sea invertebrate are found locally. Collapse has created a karst structure, with one producing a large cave within the central peak.
   Latitude 36^17' North; Longitude 85^40' West

N23   Glasford Crater is in Illinois buried under the town of the same name, near Peoria. It is 2.5 miles in diameter, and less than 430 million years old.
   Latitude 40^36' North; Longitude 89^47' West

N24   Glover Bluff Crater is in Wisconsin four miles south of the town of Coloma. Its diameter is five miles, and age is under 500 million years. The crater is mostly buried under a glacial moraine, and is being quarried, so everything from the impacting object to shatter cones are gradually vanishing.
   Latitude 43^58' North; Longitude 89^32' West

N25   Gow Crater is in Saskatchewan, forming a lake of the same name. It is 3.1 miles in diameter, and less than 250 million years old. The lake has an island that is the central peak.
   Latitude 56^27'24" North; Longitude 104^28'53" West

N26   Haughton Crater is on Devon Island in Canada's Nunavut Territory. Devon is the world's largest uninhabited island, although today several groups have scientific bases there, working on the presumption that Devon is Earth's closest analog to the environment on Mars. The crater is 14 miles in diameter, and 39 million years old. The closest inhabited village is Resolute, on the island of Cornwallis 110 miles west.
   Latitude 75^22' North; Longitude 89^41' West

N27   Haviland Crater is in Kansas between Wichita and Dodge City. With a diameter of about 45 feet, and an age of under one thousand years, it is one of the smallest and youngest craters known. Over 15,000 pounds of stony iron meteorites, rich in olivine, have been recovered from the site. Some are displayed in the Kansas Meteorite Museum and Nature Center at the crater, while others are displayed at many museums around the USA.
   Latitude 37^34'57" North; Longitude 99^09'49" West

N28   Holleford Crater is in Ontario 16.5 miles north of Kingston. It is privately owned. The diameter is 1.4 miles, and age is 550 million years. This crater is mostly buried, and was first detected during an aerial survey of the region. It can be accessed from highway 38, which has an historical marker near the center. In the remote past the crater held a lake, but today it is merely a bog.
   Latitude 44^27'28" North; Longitude 76^38'00" West

N29   Ile Rouleau Crater is in Quebec on the island of the same name in Lake Mistassini, where it forms a bay near the south end of the lake. The crater is 2.4 miles in diameter (part under water), and less than 300 million years old. Shatter cones and coesite have been found. The island is most easily reached by canoe from the town of Baie-du-Poste.
   Latitude 50^41' North; Longitude 73^53' West

N30   Jeptha Knob Crater is in Kentucky near the town of Shelbyville. It is 3 miles in diameter, and 425 million years old.
   Latitude 38^10'38" North; Longitude 85^06'41" West

N31   Johnsonville Crater is in South Carolina at the junction of the Lynches and Pee Dee Rivers. Its diameter is 8 miles, and it is less than 50 million years old.
   Latitude 33^49' North; Longitude 79^22' West

N32   Kentland Crater is in the town of Kentland, Indiana. I easily drove to it from a relative's home in Evansville on my way to Chicago. The crater is 8.3 miles in diameter, and less than 97 million years old. Erosion has left the rim little more than a bump in the road. The crater has been used as a quarry since 1880 right up to today, but coesite and shatter cones can still be found.
   Latitude 40^45' North; Longitude 87^24' West

N33   La Moinerie Crater is in Quebec's Riviere-Koksoak district. The diameter is five miles, and it is about 400 million years old. It contains a lake, Lac La Moinerie, which has some islands making a central peak.
   Latitude 57^24' North; Longitude 66^37' West

N34   Manicouagan Crater is a ring shaped lake in Quebec 190 miles north of the town of Baie-Comeau, with Route 389 from there going along the crater's eastern edge. Note: north of here the road is of poor quality. The annular lake defines the inner ring of the crater, and is used as a reservoir for a hydropower plant. The current state of the crater has the lake with a diameter of 53 miles, but it is believed to have originally been far larger. It is 214 million years old.
   Latitude 51^23' North; Longitude 61^42' West

Manicouagan Crater showing the ring shaped lake forming a reservoir. NASA photo from 2001.

N35   Maniitsoq Crater is on the west coast of Greenland, at the Inuit village of the same name. The diameter is 62 miles, and it is believed to be 3 billion years old, making it both one of the largest and oldest known craters. It was recognized as a crater in July 2012. The town can be reached by Air Greenland or by the Umiaq ferry.
   Latitude 65^25' North; Longitude 52^54' West

N36   Manson Crater is in Iowa near its namesake town. Although buried, it is 31 miles in diameter, and 74 million years old. This makes it too old to have wiped out the dinosaurs, although its landing in a shallow sea must have been an environmental catastrophe for whatever was living within a couple hundred miles of the impact. It was identified as a crater in 1959 from coesite. The impacting object was stony.
   Latitude 42^35' North; Longitude 94^33' West

N37   Maple Creek Crater is in Saskatchewan, near its namesake town. The diameter is 3.8 miles, and it is less than 75 million years old. It is buried.
   Latitude 49^48' North; Longitude 109^06' West

N38   Marquez Crater is in Texas, located under the town it is named for. It is 7.7 miles in diameter, and 58 million years old.
   Latitude 31^17' North; Longitude 96^18' West

N39   Middlesboro Crater is in Kentucky, with the town of Middlesborough located inside it. Route U.S. 25E provides access to both the town and the crater. The crater is one of the four components of the Cumberland Gap, the historic access route through the mountains. The crater is 3.7 miles in diameter, and less than 300 million years old. The world's only coal mine in a crater was located here, but has shut down. Shatter cones and coesite have been found.
   Latitude 36^37' North; Longitude 83^44' West

Roadside Plaque identifying the Middlesboro Crater. Photo courtesy of S. L. J. Russo, Planetarium Director and instructor at Big Sandy Community and Technical College (and my former student from the 1970s).

N40   Mistastin Crater is in Labrador. The diameter is 17 miles, and age is 36.4 million years. Glacial periods have effectively smoothed out the area. Today this is a lake 9.9 miles across, with an island forming a central peak 430 feet high. The crater rim allows the lake surface to be about 1000 feet above sea level. The impacting object seems to have been a siderolite. Shatter cones have been found. It was identified as a crater in 1968.
   Latitude 55^53' North; Longitude 63^18' West

N41   Montagnais Crater is in the continental shelf 125 miles south east of Shag Bay (Halifax), Nova Scotia. It has a diameter of 28 miles, and is 50.5 million years old. It was first recognized in 1987. Drilling indicated that the impacting object was stony.
   Latitude 42^53' North; Longitude 64^13' West

N42   Newporte Crater is in North Dakota, west of the town of Sherwood. It has a 2 mile diameter, and is less than 500 million years old.
   Latitude 48^58' North; Longitude 101^58' West

N43   New Quebec Crater is also known as Chubb Crater for its discoverer and more recently by the name first used by Native Americans in the region, Pingualuit Crater. It is in the far north of the Ungava Peninsula, and contains Pingualuk Lake. The crater is 2.14 miles in diameter, and 1.4 million years old. The lake is 876 feet deep, with a 520 foot high rim. It is ice covered November to July, and there are only twenty days a year of "warm" weather. Swimming is further discouraged by the inner slope down to the water being a 35 to 40 degree angle. The impacting object was stony iron. The nearest village is Kangiqsujuaq (try getting that one past a proofreader!), 62 miles to the west. The crater and surrounding area was made a National Park in 2004.
   Latitude 61^17' North; Longitude 73^40' West

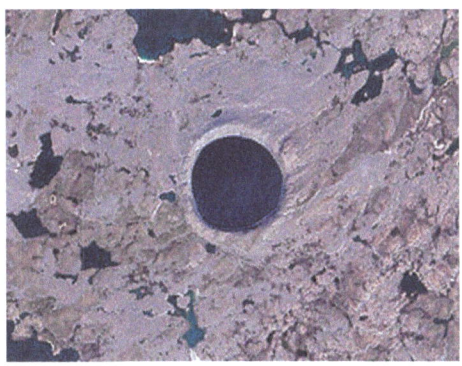

NASA photo of Chubb Crater taken from an aircraft looking southeast.

N44   Nicholson Crater is in Canada's North West Territory. It is near Dubaunt Lake, with the town of Reliance to the west. This buried crater is about 7.7 miles in diameter and under 400 million years old. It was discovered through drilling for oil.
   Latitude 62^40' North; Longitude 102^41' West

N45   Odessa Crater is in Texas, five miles southwest of the city of the same name. It is easily accessed on I-20 at exit 108. It is less than 63,500 years old. It was first recognized as a crater in 1922. There are five craters in this complex, with two of the smallest buried under wind blown soil. The largest is 550 feet in diameter with a rim five to seven feet high. It is estimated that when formed the main crater was 100 feet deep, but today is just 15 feet deep. A hiking trail is marked through the crater. More than 1500 siderolite meteorite fragments have been found, with many on display in a museum in the town of Odessa.
   Latitude 31^45'22" North; Longitude 102^28'44" West

N46   Panther Valley Crater is in New York State. It is 6 miles in diameter and 375 million years old. It is accessed from Route 28. Coesite and shatter cones have been found. It is bordered by Esopus Creek. The nearest town is Johnstown, but the major resort area around Saratoga Springs is convenient. It is believed to have impacted a shallow sea.
   Latitude 42^03'23" North; Longitude 74^23'42" West

N47   Pilot Crater is in Canada's North West Territory, about 45 miles northwest of Fond du Lac and 35 miles from Fort Smith. It is about 3.6 miles in diameter, and 445 million years old. Pilot Lake within the crater has lake trout, pike, and other popular species for fishing. The name comes from its use by airline pilots for navigational purposes. It was recognized as a crater in 1965.
    Latitude 60^17' North; Longitude 111^01' West

N48   Presqu'ile Crater is in Quebec, two miles south of Chapais. It was originally 16.5 miles in diameter, but in the past nearly 500 million years geologic activity has shrunk it. Shatter cones have been found.
    Latitude 49^43' North; Longitude 74^48' West

N49   Red Wing Crater is in North Dakota 15 miles southwest of Wafford City. It is 5.6 miles in diameter, and its dating at 455 million years old is based on findings of fossils. It was discovered in 1996 while drilling for oil.
    Latitude 47^36' North; Longitude 103^33' West

N50   Rock Elm Crater is in Wisconsin, adjacent to the town whose name it shares. The crater is 3.7 miles in diameter, and less than 505 million years old. It may be part of a group of four craters formed in the same event, including Ames Crater (N1), Decorah Crater (N17), and Slate Islands Crater (N56).
    Latitude 44^43' North; Longitude 92^14' West

N51   Saint Martin Crater is in Manitoba, north of the town of Grahamdale, and contains a lake sharing the crater's name. The crater is 25 miles in diameter, and 220 million years old. The area is covered with many glacial deposits, making the crater hard to recognize.
    Latitude 51^47' North; Longitude 98^32' West

N52   Santa Fe Crater is in New Mexico (not to be confused with a Martian crater of the same name). It is buried, but has a diameter of 9 miles, and is estimated to be under 1.2 billion years old. It is near the Chaco Canyon National Monument.
    Latitude 35^45' North; Longitude 105^56' West

N53   Serpent Mound Crater is in Ohio, northwest of the town of Peebles. It is five miles in diameter, and less than 320 million years old. Shatter cones have been found. Serpent Mound is a famous Native American construct from roughly 1000 years ago, and the combination of crater and Native materials are emphasized in local museums.
    Latitude 39^02' North; 83^24' West

N54   Sierra Madera Crater is in Texas. It can be reached by U.S. Highway 395 from the town of Midland. This crater is privately owned, being on the La Escalera Ranch. A highway marker is near the crater center, which has a central peak. The crater is nine miles in diameter, and less than 100 million years old.
    Latitude 30^36' North; Longitude 102^55' West

N55   Skeleton Lake Crater is in Ontario, 11 miles west of Huntsville, and about a three hour drive from Toronto, from where after visiting the McLaughlin Planetarium I drove to the crater. Local geological activity has warped its shape to being 5.2 by 3.0 miles. It is about 800 million years old. The south shore of the lake has a bungalow colony (no one was aware of it being a crater lake when I was there), as well as several children's summer camps run by various church groups. Swimming, boating and fishing for bass and trout are all activities available at the bungalow colony, which has a general store.
   Latitude 45^15'02" North; Longitude 79^27'01" West

N56   Slate Islands Crater is in Ontario six miles south of the town of Terrace Bay. It is 19 miles in diameter, and 450 million years old. Two main islands, Patterson and Mortimer, five minor islands, and many islets define the crater's outline in Lake Superior. Shatter cones have been found on the islands and nearby shore. The area was named a park in 1985. This crater may be part of a group of four craters all formed at the same time, including Ames Crater (N1), Decorah Crater (N17), and Rock Elm Crater (N50).
   Latitude 48^40' North; Longitude 87^00' West

N57   Steen River Crater is in Alberta seventy miles north of Fort Vermilion, sixty miles east of the town of High Level, and close to Wood Buffalo National Park. It is 16 miles in diameter, and 91 million years old. It was discovered during oil exploration.
   Latitude 59^30' North; Longitude 117^38' West

N58   Sudbury Crater is in Ontario, with the city of Sudbury inside it. The crater is believed to have originally been 160 miles in diameter, and rubble from its formation 1.849 billion years ago has been found 500 miles away in Minnesota. This makes it both the oldest and largest known crater in North America. On its northeastern rim is a totally unrelated and much younger crater (N62), the only known case on Earth of overlapping but unrelated craters. Today, thanks to glaciers, tectonic motion and other events it is 39 by 19 miles. Road 35 (Elm Street) and Highway 144 pass through much of the crater. It was recognized as a crater in 1970, but was actively mined well before then. Shatter cones have been found. Mines are the largest producer of nickel in the Western Hemisphere, and other metals mined include palladium and platinum. Deep within a no longer active part of the mine is a neutrino observatory, which figured in an excellent science fiction novel, *Hominids*, (and two sequels) by Robert J. Sawyer. It won a Hugo as best SF novel of 2003.
   Latitude 96^36' North; Longitude 81^11'

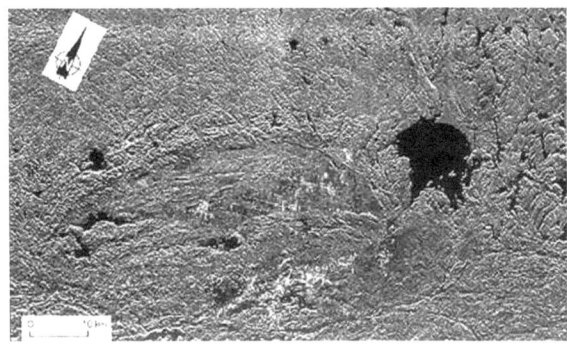

**NASA infra-red photo from the Space Shuttle Challenger in 1984.
Sudbury is the elongated ellipse, with the water-filled Wanapitei Crater (N62) to the upper right.**

N59   Upheaval Dome Crater is in Utah in Canyonlands National Park, 42 miles from the town of Moab. The crater has three concentric rings with a diameter of 6.2 miles. It is less than 170 million years old. It was identified in 2008 following the discovery of coesite.
   Latitude 38^26' North; Longitude 109^54' West

N60   Tunninik Crater is on Victoria Island in Canada's North West Territories. It originally was named Prince Albert Crater for the peninsula it is on. It is 16 miles in diameter, and under 350 million years old. Shatter cones have been found. It was identified in 2010.
   Latitude 72^28' North; Longitude 113^58' West

N61   Viewfield Crater is in Saskatchewan south of the town of Stoughton. It is 1.5 miles in diameter, and 190 million years old. It as discovered in 1972 in the course of drilling for oil, and the broken up rocks beneath the crater have concentrated enough oil that 400 barrels per day are pumped from around the crater's edges.
   Latitude 49^35' North; Longitude 103^04' West

N62   Wanapitei Crater is in Ontario at the town of Wahnapitae, on the edge of the unrelated Sudbury Crater (N58). It is 4.5 miles in diameter and only 37.2 million years old, about 5% the age of the adjoining Sudbury Crater. Local geological activity has left it 10.3 by 8.7 miles. Associated coesite has been found. This crater is included in the photograph of Sudbury, N58.
   Latitude 46^45' North; Longitude 80^45' West

N63   Weaubeau Crater is in Missouri near the town of Osceola. The diameter is twelve miles, and it is 330 million years old.
   Latitude 37^59' North; Longitude 93^38' West

N64   Wells Creek Crater is in Tennessee. It can be reached going north on Route 149 from the town of Pollard, and Cumberland is inside the crater. The diameter is 7.46 miles, and it is 200 million years old. Shatter cones have been found.
   Latitude 36^22'40" North; Longitude 87^39'30" West

N65   West Hawk Crater is in Manitoba buried beneath a lake of the same name in White Shell Provincial Park sixty miles east of Winnipeg. The crater is much smaller than the overlying lake, just 1.4 miles in diameter, and 351 million years old.
   Latitude 49^46' North; Longitude 95^11' West

N66   Wetumpka Crater is in Alabama twelve miles north of Montgomery, and with a town sharing its name on its western edge. The crater is 4 miles in diameter, and 81 million years old. The Coosa River forms the western edge of the crater. The impact occurred at a time when this was a coastal region, and this crater is said to be the best preserved of any shallow marine impact. Four small ponds are scattered around the inside of the crater. Several local roads go through the crater, including U.S. 231, Trotters Trail, and Buck Ridge Road. The crater was discovered in 1969, and fossils have aided its dating. Coesite and enhanced iridium have been found. The town holds an annual "craterfest" around the third weekend in April.
   Latitude 32^31' North; Longitude 86^10' West

N67   Whitecourt Crater is in Alberta, six miles from the town of the same name. It is only 118 feet in diameter, 30 feet deep, and less than 1100 years old. While long known to hunters as an odd depression in a dense pine forest, it was first identified as a meteor crater in 2007. The impacting object was a siderolite, based on more than 3000 fragments collected in the area. The Alberta Government has declared a preservation zone 20 meters wide around the crater to prevent looters from destroying it and walking off with all the meteorites. Penalties range up to a year in prison and a hefty fine for violations.

Latitude 54^00' North; Longitude 115^36' West

OCEANA

O1   <u>Baguio</u> Crater is in The Philippines, in the port area known as La Union of the Luzon city of Baguio. It extends under water in the harbor, and has had little study.
    Latitude 16^23' North; Longitude 120^26' East

SOUTH AMERICA

S1   Araguainha Crater is in Brazil, on the border of the Mato Grasso and Goias states. It is between a village sharing its name and Porta Branca. The diameter is 25 miles, and it is 354 million years old. It was recognized in 1973 thanks to the presence of shatter cones and coesite. A river of the same name crosses the crater, and road MT-306 provides access. The impacting object was probably a siderolite.
   Latitude 16^47' South; Longitude 52^59' West

S2   Campo del Cielo Craters are in Argentina. The nearest large city is the capitol of Chaco Province, Resistencia. There are 26 craters, the largest 350 by 300 feet across, with siderolite fragments found out to a distance of 37 miles. Carbon dating of charred wood shows they are 2400 years old. Analysis has found the impacting object was 92.9% iron, 6.7% nickel, 0.4% carbon. Local native tribes made weapons from fragments, which quickly attracted Spanish attention in the Sixteenth Century. However, interest waned, and the area was pretty much ignored until the late Twentieth Century.
   Latitude 27^38' South; Longitude 61^42' West

S3   Carancas Crater is in Peru. The impacting chondrite was seen to fall September 15, 2007, leaving a 40 foot wide, 15 foot deep hole. Steam and gas from this hole made those who initially approached it ill, apparently from well known deposits of arsenic in the local soil, although troilite found in the meteorite has also been blamed. It fell next to a village of the same name, south of Lake Titicaca. This crater has one of the highest altitudes of any known crater, having landed on a plateau 12,400 feet above sea level.
   Latitude 16^39'52" South; Longitude 69^02'38" West

S4   Colonia Crater is in Brazil, located in the Sao Paolo neighborhood of the same name. It has a 2 mile diameter, and is less than 36 million years old. In the past it contained a lake, but this has long since vanished, although some of the crater's interior remains swampy.
   Latitude 23^52'15" South; Longitude 46^42'30" West

S5   Monpuraqui Crater is in Chile, south of the town of Peine, and near San Pedro de Atacama. The diameter is 1210 feet, and it is 660,000 years old. The crater is 112 feet deep. It was first recognized in 1966. The impacting object was a siderolite which came in from the northwest. It is at an altitude of 9200 feet. A small seasonal lake forms in the northern part of the crater.
   Latitude 23^55'41" South; Longitude 68^15'42" West

S6   Riachao Ring is in Brazil about 65 miles from the nearest large town of Carolina. It is 2.8 miles in diameter, and under 200 million years old. This is a complex of ten craters, the largest within the ring being the Northern Basin. It lies seven miles north of the main complex, and has a diameter of 2700 feet. The next largest are 2200 and 700 feet in diameter. It is believed an object coming in from the north exploded just before impact, forming the multiple craters.
   Latitude 07^43' South; Longitude 46^39' West

S7   Rio Cuarto Crater is in Argentina near the town of the same name. It is 2.8 miles in diameter, and 10,000 years old. It as identified in 1990 as resulting from a grazing impact by a chondrite that came in at a low angle of fifteen degrees, creating an elliptical crater.
   Latitude 32^52'40" South; Longitude 64^14'25" West

S8   Serra da Cangalha Crater is in Brazil near the town of Pedro Afonso. It is 8 miles in diameter and 220 million years old. It has a series of nested rings, 7, 3.5, and 1.8 miles in diameter. It was recognized in 1973 thanks to the presence of shatter cones and coesite.
   Latitude 08^05' South; Longitude 52^07' West

7.5 mile diameter Serra da Camgalha Crater from space.

S9   Vargeao Crater is in Brazil, with a town named Vergeao inside the southern part of the 8 mile diameter crater. The crater is less than 70 million years old. The town is a lot younger than that. Shatter cones and coesite have been found.
   Latitude 26^50' South; Longitude 52^07' West

S10   Vista Alegre Crater is in Brazil near the town of Porto Umiao. It is 6 miles in diameter and less than 65 million years old.
   Latitude 25^57' South; Longitude 52^41' West

CRATERS PER MILLION SQUARE MILES

Place	Number of Craters	Square Miles* (in millions)	Ratio: craters/million square miles
Africa	20	11.7	1.6
Asia	23	17.2	1.3
Australia	29	2.97	10
Antarctica	3	5.4	0.58
Europe	42	3.93	10.7
North America	67	9.54	7.2
South America	10	6.89	1.4
The Moon	~600,000	23.5	27,000**

*Millions of square miles in surface area
**Limited to craters with diameters greater than 0.62 mile

PLANET MERCURY

M1 Caloris Basin is one of the largest impact sites in the Solar System. It is about 1300 miles across, and has many smaller craters within it. Counts of such internal features suggest an age of about 3.8 billion years.

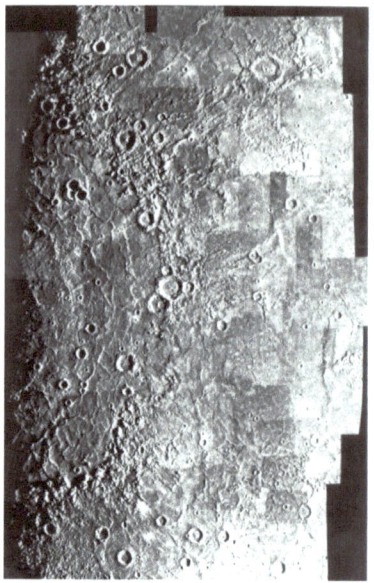

Caloris Basin on Mercury from Mariner 9.

M2 Mercury's surface shows large numbers of impact caters.

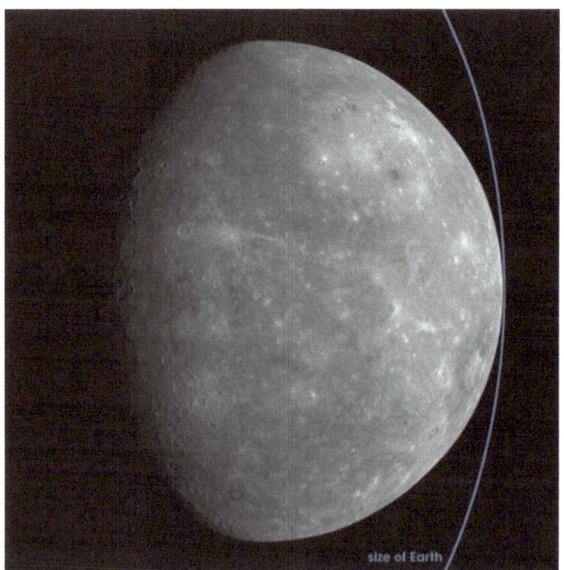 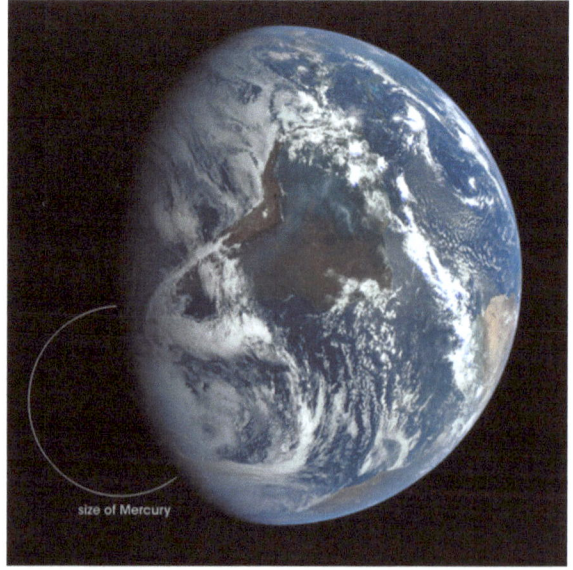

Mercury's surface is near saturation. Note outline showing Earth's size compared to Mercury.

OUR MOON

L1 Copernicus Crater at the top of this Apollo 17 photograph is 58 miles in diameter, and 700 million years old.
   Latitude 10^ North; Longitude 20^ West

Crater Copernicus from lunar orbit.

MARS

R1 <u>Hellas</u> Basin is the largest crater on Mars, about 800 miles in diameter, and forming a deep depression in the planet. It shows considerable evidence of various forms of erosion, including suggestions of some by flowing water.

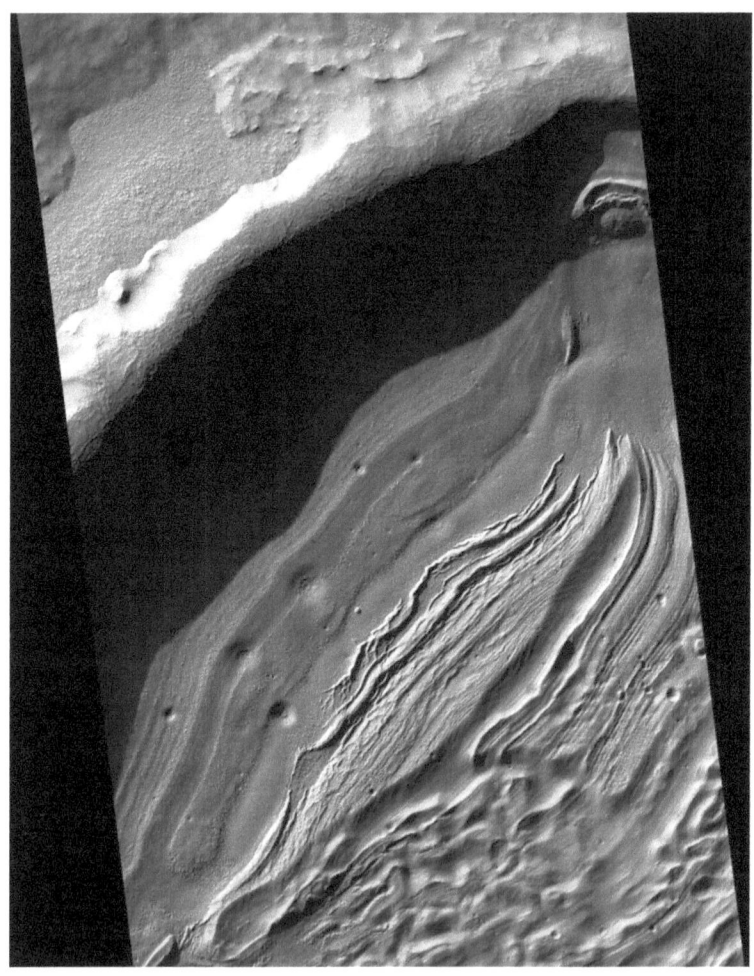

A small section of Hellas Basin, where something larger than
either of Mars' moons hit the planet, from the Mars Reconnaissance Orbiter.

R2   Some Martian craters show evidence of water forming now vanished ponds within them, much as we also find on Earth.

Landslides near a fresh (2 year old) crater on Mars.

CALLISTO

T1   <u>Valhalla</u> Basin, showing the concentric ring structure formed as the crater walls slumped under gravity's attraction. This crater is about 1100 miles across.

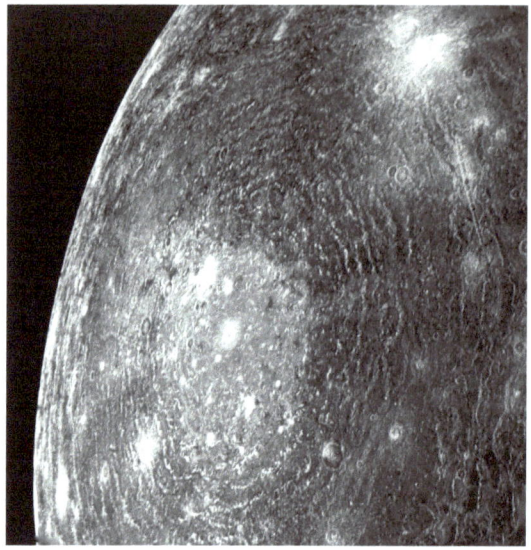

Valhalla Basin, largest crater on Callisto.

T2   Callisto's surface is saturated with craters, meaning that for every new crater formed, an old one is destroyed. This suggests Callisto has an extremely old surface, perhaps going back 4 billion years.

Callisto's surface is saturated with craters.

MIMAS

Mimas is a moon of Saturn.

I1 <u>Herschel</u> Crater is unusual, because it is believed the central peak is actually part of the impacting object sticking out of the crater. Its jagged nature, as opposed to the usual appearance of central peaks, seems to prove this. The impact must have been at very slow relative speeds. This shows the crater from the Cassini spacecraft in 2005. The diameter is 86 miles, it is 6 to 8 miles deep, with the central peak extending 4.5 miles above the floor. It is 4.1 billion years old. It was discovered in 1980. There are also craters named Herschel on our Moon and elsewhere.
  Latitude 0^; Longitude 100^

Herschel Crater on Mimas in 2005.

INDEXES

Tourist Facilities

Warning: this list may become out of date with regard to any particular site. The definition of "facilities" ranges from as little as a hiking trail to hotels, museums, restaurants, and shopping malls. There is no assurance as to what languages may be spoken at many foreign locations. Check what is available before going to any site, and be aware of the local climate. Even in North America or Europe you may need to take precautions regarding clothing, availability of food and other supplies, and housing.

A4	A17*	A18*
B11	B21	B22
D14	E3	E5
E21	E27	N3
N5	N9	N17
N23	N24	N27*
N30	N32	N36
N38	N39*	N45*
N46	N53	N55
N58*	N62	N64
N66	S1	S4
S9		
D6*		

*indicates related museum

National or Other Parks; World Heritage Site

A17	D7	D19	D21
N20	N39	N52	N56
N57	N59	N65	N67

Privately Owned Craters

A3, A6, D3, N3, N28, N54

Crater Pairs

A3, D25, E7/9,
A20, E11/N11?
E20/23
E31/35
E36, N12

Crater Chains

N1, 17, 50, 56

Clusters of Craters

B12, B13, B20, D14,
E15, E27, N45, S2, S6,
B23

Crater Ages

Under 100,000 years	*100,000 to 1 million years*	*1 to 5 million years*
A1	A2	A3
A7	A6	A5
A15	A18	A12
B12	B11	A13
B17	D6	B2
B19	D7	B6
B20	S5	B9
B21	N6	
D3	N43	
D14		
D15		
D26		
D27		
E12		
E15		
E27		
N3		
N27		
N45		
N67		
S2		
S3		
S7		
B23		

10 to 50 million years	*50 to 100 million years*	*100 to 200 million years*
B1	A11	A9
B3	A15	A10
B7	A19	A14
B10	B4	B18
B13	B8	D12
B14	B15	D18
B16	D4	D20
D5	E2	D25
D8	E3	E24
D11	E19	E26
E1	E42	E29
E7	N11	E40
E9	N18	E42
E31	N20	N8
E34	N32	N13
E35	N36	N59

10 to 50 million years
N2
N10
N31
N40
N62
S4
S10

50 to 100 million years
N37
N38
N54
N57
N66
S9

100 to 200 million years (cont.)
N61
A20

200 to 350 million years
E4
E17
E30
E33
E37
N9
N12
N15
N16
N19
N25
N29
N39
N51
N53
N60
N63
S8
N23
N24
N30
N33
N34
N42
N44
N46
N47
N48
N49
N56
N65
S1

350 to 500 million years
C3
D21
D28
E6
E10
E13
E16
E20
E23
E28
E38
N1
N5
N7
N14
N17
N21
N22

500 million to 1 billion years
A8
D1
D9
D10
D16
D17
D23
D24
E5
E8
E25
E32
E36
N4
N28
N50
N55
L1

Craters More Than One Billion Years Old

A17	B5	D2
D13	D19	D22
D29	E14	E18
E21	E22	E39
N35	N52	N58
M1	I1	A21

Quarries or Mines Associated with Impact Craters

A12	N8	N24
N32	N39	N58

Impact Craters that Contain Related Lakes Today

A4	A18	B6
B9	B11	B12
D1	E3	E15
E18	E19	E24
E30	E36	N5
N14	N18	N25
N33	N40	N43
N47	N51	N64
N66	S5	

Impact Craters By Size (Diameter)

Less Than One Mile Diameter		*1 to 3 Mile Diameter*		*3 to 5 Mile*
A1		A2	A6	A4
A3		A12	A13	A5
A7		A13	A16	B5
A14	A15	B1	B14	
A18	A19	B2	B22	
B12	B11	B3	D4	
B17	B16	B4	D5	
B20	B18	B7	D8	
D3		B19	D9	D23
D6		B21	D12	D24
D7		D13	D16	E8
D14	D20	D19	E10	
D15	E4	D21	E21	
D18	E6	E5	E22	
D26	E7	E13	E24	
D27	E14	E16	E37	

Less Than One Mile Diameter		1 to 3 Mile Diameter		3 to 5 Mile (cont.)
E12	E32	E20	E39	
E17	E35	E25	N1	
E23	E36	E28	N7	
E27	E38	E40	N18	
N3		N5	N2	N19
N27	N6	N13	N31	
N45	N22	N14	N32	
N67	N23	N15	N38	
S2		N28	N16	N44
S3		N29	N17	N46
S5		N42	N21	N49
B23	N43	N24	N52	
N61	N25	N54		
N65	N30	N55		
S4	N33	N59		
S6	N35	S8		
S7	N39	S9		
A21	N47	S10		
N50	A20			
N53				
N62				
N66				
A20				
D11				

10 to 50 Miles Diameter	50 to 150 Miles Diameter	More Than 150 Miles Diameter
A9	B8	A17
A10	B13	B15
B6	C1	C2
B9	N10	C3
B10	N34	N11
D1	N35	N58
D2	I1	M1
D5	L1	R1
D10	T1	
D11		
D17		
D22		
D25		
D28		
D29		
E1		
E2		

10 to 50 Miles Diameter (cont.)
E3
E9
E18
E19
E26
E29
E31
E33
E34
N4
N8
N9
N12
N36
N40
N48
N51
N56
N57
N60
N63
S1

Nickel-Iron	Chondrites	Achondrites	Stony-Iron
A7	A9	D24	D6
B12	A18	N27	
B15	B13	N43	
B20	D1		
D3	N1		
D17	N9		
D26	N11		
D27	N36		
E27	N41		
N3	S3		
N40	S7		
N45			
N58			
N67			
S1			
S2			
S5			
B23			

ALPHABETICAL LIST OF ALL NAMES

A
Acraman D1
Adelaide (City) D5, D8
Ajana D28
Alabama N64
Aland Island E22
Alaska N2
Alberta N20, N57, N67
Algeria A1, A11, A13, A16
Alice Springs D3, D12, D14
Amelia Lake D2
Ames N1
Amguid A1
Aorounga A2
Aouellol A3
Aragon E34
Araguianha S1
Aral Sea B3
Argentina S2, S7
Arizona N3
Arkenu A20
Arnhem Land D13, D18
Atlantic City N10
Avak N2
Azuara E1

C
Caloris M1
Calvin N7
Campo del Cielo S2
Canada N9, N12, N14, N20, N21, N25, N26, N28, N29, N33, N34, N37, N40, N41, N43, N44, N48, N51, N55, N56, N57, N58, N60, N61, N62, N65
Canyonlands N59
Carancas S3
Carswell N8
Carolina S6

B
Baden-Wurtemberg E35
Baguio O1
Baie Comeau N34
Baie du Poste N29
Barents Sea E26
Barringer N3
Barrow (town) N2
Bavaria E31
Beaverhead N4
Belarus E21
Belaya River B6
Beyenchime B1
Bigach B2
Bjornoya Island E26
Bobrinets E42
Boltysh E2
Borroloola D9
Bosumtwi A4
Bowers C1
Boxhole D3
Brazil S1, S4, S6, S8, S9, S10
Brent N5
Bushy Creek N6

D
Dalgaranga E4
Darwin D7
Deacon N5
Decorah N17
Dell N4
Deep Bay N18
Dellen E3
Des Plaines N19
Devon Island N26
Dhala B5

51

Carswell N8
Chaco Canyon N52
Chaco Province S2
Chad A2, A5
Chapais N48
Charlevoix N9
Chesapeake Bay N10
Chicxulub N11
Chile S5
Chubb N43
Chukcha B4
Chuktoka B6
Clearwater Lakes N12
Cloud Creek N13
Coloma N24
Colonia S4
Congo A8, A19
Connolly Basin D4
Coosa River N66
Copernicus L1
Couture N14
Crooked Creek N15
Cumberland N64
Cypress Hills N20

F
Finland E14, E17, E18, E22, E30, E36
Flagstaff N3
Flaxman D8
Flinders Mount D1
Flynn Creek N22
Fond du Lac N47
Foelsche D9
Fort McKenzie N12
Fort Smith N47
Fort Vermilion N57
France E33
Franklin-Gordon D7

H
Haapaselka E36
Halifax N41
Halls Creek D11, D23

Dirham E28
Dobele E4
Dodge City N27
Dornogovi B18
Douglas River N8
Drevan River E24
Dubaunt Lake N44
Durham Downs D25

E
Eagle Butte N20
Eastern Cape A6
Eastern Kasai A19
Eboulements N9
Egypt A7
Elbow Creek N21
El'Gygytgyn B6
Ennedy A2
Esopus Creek N46
Estonia E12, E15, E16, E29

G
Gainesville N22
Gardnos E5
Germany E31, E35
Ghana A4
Gibb River Road D23
Gibson Desert D4, D26
Gilmour Lake N5
Glasford N23
Glenwood N7
Glikson D10
Globino E37
Glover Bluff N24
Goat Paddock D11
Goias S1
Gosses Bluff D12
Gow N25
Goyder D13
Grahamdale N51

Hassi Delaa A13
Haughton N26
Haute Vienne E33
Haviland N27
Hellas R1
Henbury D14
Herschel I1
Hickman D15
High Level N57
Hiiumaa E16
Holleford N28

I

Iceland E11
Ile Rouleau N29
Illinois N19, N23
Ilumetsa E12
Ilyinets E13
India B5, B11, B15
Indiana N32
Iowa N17, N36
Iraq B19
Iso-Naakkima E14

L

La Escalera N54
La Judie E33
La Loche N8
La Moinerie N33
Labrador N40
Lahojsk E21
Lake Diefenbaker N21
Lake Mistassini N29
Lake Teague D22
Lappajarvi E19
Latvia E4
Lawn Hill D17
Lebanon, Mo N16
Libya A10, A20
Lithuania E25, E40
Little Sandy Desert D10
Liverpool, Aust. D18

Granby E6
Grant N4
Greenland E26, N35
Greensburg N6
Gusev E7
Gweni-Fada A5

J

Janisjarvi E8
Jebel Waqf Suwwan B7
Jeptha Knob N30
Johnsonville N31
Johnstown N46
Jordan B7
Judbarra/Gregory D19

K

Kaalijarv E15
Kansas N27
Kalahari A9
Kallkop A6
Kaluga E10
Kamensk E9
Kamil A7
Kara B8
Kara-Kul B9
Kardla E16
Karikkoselka E17
Karnansaari E19
Katherine D19
Kazakhstan B2, B3, B16, B22
Kelly West D16
Kentland N32
Kentucky N30, N39
Keurusselka E18
Keuruu E18
Kgagodi A22
Khatanga B13
Kiev E29

Lockne E20
Logancha B10
Logoisk E21
Lonar B11
Louisiana N6
Lugove E13
Luizi A8
Lumparn E22
Luzon O1
Lynches River N31

M
Ma'an B7
Macha B12
Madhya Pradesh B5
Maharashtra B11
Malingen E23
Manicouagan N34
Maniitsoq N35
Manitoba N51, N64
Manson N36
Mantta E18
Maple Creek N37
Marcellus N7
Marquez N38
Matt Wilson D19
Matto Grasso S1
Mauritania A3, A14, A16
Maysan B19
Medicine Hat N20
Meekathara D29
Mexico N11
Michigan N7
Middlesboro N39
Mien E24
Midland N54
Missouri N15, N16, N63
Mistastin N40
Mizarai E25
Mjolnar E26
Moab N59
Mongolia B18
Monpuraqui S5

Kirovohrad E2
Kingston N28
Krivoy E42
Kumasi A4
Kuukkarinselka E36

N
Namibia A12
Nasviken E3
Nes E5
Neugrund E28
New Mexico N52
New Quebec N43
New York State N46
Newporte N42
Newsom D15
Nicholson N44
Nordlingen E31
North Dakota N42, N49
Northern Territory D2, D13, D14, D16, D18, D19, D24
Northwest Territory N44, N47, N60
Norway E5, E26, E32
Nova Scotia N41
Novopavlovka B2
Nunavut N26

O
Oasis A10
Obolon E29
Odessa N45
Osceola N63
Ohio N53
Ontario N5, N29, N55, N56, N58, N62
Oodnadatta D20
Oranjemund A12
Orion N20
Osmussaar E28

Montagnais N41
Montgomery N66
Montana N4
Morasko E27
Moravia E31
Morokweng A9
Mortimer N56
Mount Burnett D23
Mount Isa D18
Mount Magnet D6
Mount Timondina D20

P
Paasselka E30
Pamir Mountains B9
Panther Valley N46
Parys A17
Patterson N56
Pedro Afonso S8
Pee Dee River N31
Peebles N53
Peine S5
Peoria N23
Peru S3
Petajavesi E17
Philippines O1
Piccaninny D21
Piccnululu Park D21
Pieksamaki E14
Pilot N47
Pingualuit N43
Poland E27
Polava E12
Poltova E29
Pollard N64
Popigai B13
Porta Branca S1
Porto Umaio S10
Povungnituk N14
Poznan E27, E7, E8, E9, E10
Presqu' Ile N48
Pretoria A18
Prince Albert N60

Ostersund E20
Ouarkziz A11

Q
Qaraghandy B16
Quebec N9, N12, N14, N33, N34, N48
Queensland D18

R
Ragozinka B14
Red Wing N49
Reindeer Lake N18
Reliance N44
Resistenza S2
Resolute N26
Riachao S6
Ries E31
Rio Cuarto S7
Riverhurst N21
Riviere Koksoak N33
Ritland E32
Rochechouart E33
Rock Elm N50
Rogaland E32
Ross C2
Rostov E7
Roter Kamm A12
Rubielos de la Cerida E34
Russia B1, B4, B6, B10, B12, B14, B17,

S

Saaremaa E15
St. Lawrence River N9
St. Martin N51
Sakha B12
San Pedro de Atacama S5
Sandstone D29
Santa Fe N52
Sao Paolo S4
Saratoga Springs N46
Saskatchewan N8, N18, N21, N25, N37, N61
Saudi Arabia B20
Savouin E30
Scotland E39
Serpent Mound N53
Serra de Cangalha S8
Shag Bay N41
Shelbyville N30
Sherwood N42
Shiva B15
Shoemaker D22
Shunak B16
Siberia B6, B10, B12
Sichuan B21
Sierra Madera N54
Sikhote-Alin B23
Skvira E41
Skeleton Lake N55
Slate Islands N56
Sob River E13
Sobolev B17
Sobyk River E13
South Africa A6, A9, A17, A18
South Australia D1, D5, D8, D20
South Carolina N31
Spain E1, E34
Spider D23
Steen River N57
Steinheim am Albuch E35
Stoughton N61
Strangways D24
Suvasvesi E36
Svalbard Islands E26

T

Tabun-Khara B18
Taimyr B4
Tajikistan B9
Talawaya Trail D4
Talemzane A13
Talundilly D25
Tamanghasset A1
Tanami Road D27
Tasmania D7
Tecumseh N5
Temimichat A14
Tennant Creek D16
Tenoumer A15
Tennessee N22, N64
Ternovka E37
Terrace Bay N56
Texas N38, N45, N54
The Minch E39
Tiasmyn River E2
Tin Rider A16
Tindouf A11
Tinsryd E24
Tonui E16
Tookoonooka D25
Toronto N55
Tswang A18
Tunninik N60
Tvaren E38

U

Ukraine E2, E13, E29, E37, E41, E42
Ukmerge E40
Ullapool E39
Umiaq N35
Umm al Bini B19
Ungava Peninsula N43
Unia River A19
Upheaval Dome N59
Utah N59
Uweinat A7

Sweden E3, E6, E19, E20, E23, E24, E30
Sudbury N58

V

Vaal River A17
Valhalla T1
Vargeao S9
Veevers D26
Vepriai E40
Victoria Island N60
Victoria River Road D19

Viewfield N61
Vilnius E40
Vimpeli E19
Virginia N10
Vista Alegre S10
Vredevort A17

W

Wabar B20
Wafford N49
Wanapitae N62
Weaubeau N63
Wells Creek N64
Wembo-Nyama A19
West Australia D4, D6, D11, D15, D22, D23, D26, D27, D28, D29
West Hawk N65
Wetumpka N66
Whitecourt N67
White Shell Park N65
Wichita N27
Wilkes C3
Wiluna D22
Windy Corner D4
Winnipeg N65
Winslow N3
Wisconsin N24, N50
Wolfe Creek D27
Wood Buffalo Park N57
Woodleigh D28

X

Xiuyuan B21

Y

Yakutia B1
Yalgoo D6
Yarrabubbo D29
Yemen B20
Yucatan N11
Yugoneki B8

Z

Zapadnaya E41
Zaragoza E1
Zelenyga E42
Zhamanshin B22

OTHER BOOKS BY THIS AUTHOR

WORKS OF ASTRONOMY
Useful Star Names (2011), 70 pages
Our Neighbor Stars (2012), 56 pages
Moons of the Solar System (2013), 70 pages
Dwarf Planets and Asteroids: Minor Bodies of the Solar System (2014), 69 pages

SCIENCE FICTION AND FANTASY
Time for Patriots (2011), 208 pages
The Mountain of Long Eyes (2013), anthology of 28 stories

Review Requested:
If you loved this book, would you please provide a review at Amazon.com?